PATENT DRAWINGS

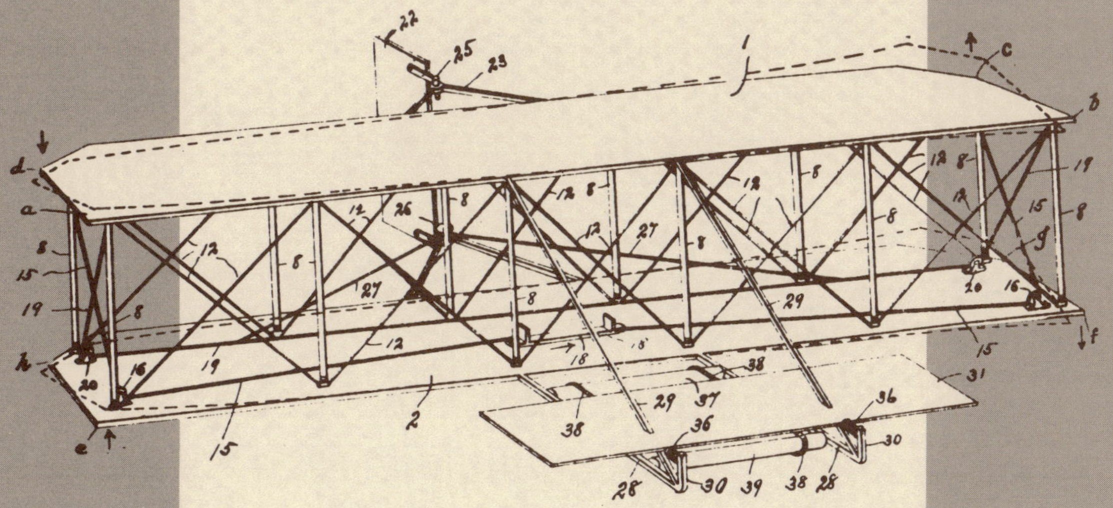

Introduction by
Kendall S. Dood

MILESTONE DOCUMENTS IN THE NATIONAL ARCHIVES

National Archives and Records Administration

Washington, DC

Published for the
National Archives and Records Administration
By the National Archives Trust Fund Board
1986

Library of Congress Cataloging-in-Publication Data

Dood, Kendall S.
Patent drawings.

(Milestone documents in the National Archives)
1. Patents — United States — Drawings. I. United States. National Archives and Records Adminstration. II. Title. III. Series.
T223.U3D66 1986 608'.022 86-16329
ISBN 0-911333-47-9

An Introduction

Drawings of inventions have been required from applicants for U.S. patents since the first U.S. patent statute was enacted in 1790. The history of patent drawings, however, goes back even further. Inventors petitioning colonial legislatures for patents were required to furnish drawings of their inventions, as were applicants to the Crown of England, where drawings were referred to in patents as early as 1673. The purpose of these requirements was in all cases the same: to ensure that the invention would be understood by all parties concerned. This included not only government officers responsible for granting patents but also other inventors, mechanics, manufacturers, patent agents, and judges and juries of courts in which legal cases involving patents might be tried — anyone, in short, who might have some interest in the information contained in the patents.

Of course, the invention was not to be understood solely from the drawing. A written description called a "specification" was also required. In fact, only a certified copy of the specification was stipulated by the 1790 patent statutes as "competent evidence" in courts where questions regarding the exact nature of the invention protected by a patent might arise. In 1836, however, the statutes were amended to include certified copies of the drawings as "competent evidence," and in 1837 they were amended again to require inventors to submit two identical sets of drawings with their patent applications, one to remain in the Patent Office as before and one to be attached to the patent issued to the inventor.

Until 1880, a model of the invention was also required as part of the patent application; often, the drawing submitted with the application was a rendering of this model. Many models were made especially to be rendered in the drawing. Of the models that still exist, those exhibiting the invention in cut-away form or in cross section are most easily recognized as draftsman's models. However, many models rendered in the drawings were constructed for other purposes: working models often served experimental or demonstration purposes, and some were actually production models, other examples of which eventually found their way into public use or sale. In many cases, patent drawings of models known to have been in actual use are the best or the only pictorial representations of them we now have.

The legal status of patent drawings as "competent evidence" after 1836 was not the only reason copies of them were required to be attached to all newly issued patents the following year. Shortly after the Patent Act of 1836 was signed into law, a fire completely destroyed the Patent Office and all the records in it. Immediately afterward, efforts were begun to replace the lost patent specifications, drawings, and models, many of which were highly valued at the time for patriotic as well as technological and other reasons. Records of patents granted before 1814 were especially prized because the Superintendent of Patents had heroically saved them from destruction by the British Army during its occupation of Washington that year. The papers accompanying Robert Fulton's application for a patent on his steamboat constitute one such set of records. The drawings, which were said to have been executed by Fulton himself, were also highly esteemed for their artistic merit.

Of the more than 10,000 sets of patent records destroyed in 1836, only about 2,800 were ever restored. The usual method for restoring the drawings was to make a copy of any certified copy of the original that could be obtained. But certified copies had been made for only a small number of the missing patents, and many of these were never found. Consequently, when such copies were not available for inventions that were considered to be particularly important, the Patent Office resorted to making new drawings from the best information it could obtain regarding the appearance of the originals.

Eli Whitney's cotton gin was one such invention, and, although a certified copy of the original drawing was later found, the Patent Office made a new drawing in 1845 based on early models of the gin and the description provided by a certified copy of the original patent specification. Although these Patent Office copies were for the most part well done, they generally reflect the drawing style of the 1840s rather than that of the originals. Moreover, errors inevitably crept into many of them. For example, a careful comparison of the 1845 cotton gin drawing with the certified copies of the original specification and drawing reveals a number of discrepancies. The lawmakers' decision in 1837 to require a copy of the drawing to be attached to each patent helped ensure that at least one copy would be available in the future if the original in the Patent Office were to be accidentally destroyed.

The statutes that required drawings to be submitted with patent applications did not specifically address style and format, but standards regarding these matters developed over the years, aided by rules promulgated from time to time by the Patent Office. Before 1836, applications received by the Patent Office were merely recorded, receiving no particular scrutiny before patents were issued for them. Hence, although there were published rules regarding patent drawings as early as 1811, the drawings were often found to be defective only after the patent had been granted. In some cases they were so defective that a drawing had to be made from the Patent Office model to replace the original drawing. The practice of reissuing patents with corrected drawings began during this period so that the patent holder's right to his invention could be salvaged when the courts held his original patent invalid because of defects in his drawing. From 1836 on, however, all patent applications were routinely examined by the Patent Office as to the clarity and accuracy of their specifications and drawings as well as for the novelty and usefulness of the inventions disclosed in them. Drawings found to be unsatisfactory were returned to the inventor for correction before the patent was issued.

Although there were various rules regarding the style and format of the drawing, there was always only one fundamental rule governing what was depicted in it: the thing pictured had to be the thing described in the patent specification. However, protection afforded by a patent was not limited to the form of the invention described in its specification and illustrated in its drawing. For an invention to be patentable, the statutes required that it be embodied in some physical form, but it was also understood that it was the invention itself that was being patented — not necessarily the particular form in which it was embodied.

The description of the invention actually protected by the patent is found in a part of the specification called the "claims," in which the invention is usually described in more general terms than in the rest of the patent. There are two consequences of this. First, one cannot automatically avoid infringing a patent just by making some minor alteration in the described embodiment, even if the alteration is itself a patentable improvement. Second, the patentee may be able to embody his invention in a variety of different physical forms without losing his patent protection. Since the breadth of a claim — the degree of generality with which it describes the invention — is restricted primarily by other existing inventions in the same field, the broadest claims are usually found in patents for inventions that open up completely new areas of technology. Bell's telephone patent is one example. Early commercial embodiments of this invention, although differing substantially in appearance from the embodiment depicted in the drawing, were still protected by his original patent.

Applicants were not limited to using the drawing merely to depict a single embodiment of their inventions. They were also permitted to show alternative embodiments as well as examples of the "prior art" — earlier attempts of others to solve the same problems. In addition, applicants were permitted to use the drawing to help explain the principles of operation of their inventions.

Patent drawings were never required to be "working drawings" such as a manufacturer might need to make the actual article or machine. Nor were they required to picture the invention to scale. It was enough if anyone skilled in the relevant field of technology could make the improved article or machine from the information in the patent drawing and specification, whether or not he also had to rely on some of his own knowledge and experience to do it.

The requirement established in 1837 for two sets of drawings was dropped in 1870 when the government began printing complete copies of patents as they were issued. Before then, copies of patent drawings had been made by hand at the expense of those who requested them. Although there was always a demand for copies of patent drawings from persons interested in the technical information they provided, handmade copies were too expensive to be disseminated widely. To help meet this demand, in the 1820s engravings of patent drawings began to appear with descriptions of newly patented inventions in journals aimed at mechanics and patent practitioners. These efforts were generally short-lived, however, because of the expense of such reproductions and because only selected patents were reproduced. The first attempt to illustrate every invention patented within a year appeared with the *Annual Report of the Commissioner of Patents to Congress* for the year 1853. An addendum to the report included brief descriptions of all inventions patented, a claim from each patent, and small woodcuts based on the original patent drawings. These woodcuts were replaced with copperplate engravings in 1854, and extra copies of the reports were printed and distributed nationwide.

The government first published photolithographs of patent drawings in 1861, when the Commissioner of Patents was authorized to print 10 copies of the specification and drawing for each newly issued patent. Publication began with patents issued March 2 of that year but ended abruptly with the issue of November 1, when all available government resources were redirected to the Civil War effort. Photolithographing of drawings was resumed in 1869 on an experimental basis, and in 1871 the government once again began publishing complete printed copies of patents as they were issued. Not surprisingly, the rules the Patent Office published that year concerning the style of the drawings submitted with applications were far more comprehensive than they had been in the past. Good quality in the reduced, printed copy and a uniform overall appearance were thus assured.

Also in 1871 the process of printing copies of the drawings of all previously issued patents began. All the old drawings — approximately 100,000 of them — were traced in 7 years, and photolithographing of the tracings was completed shortly after that. The printing of the old patent specifications, however, proceeded at a much slower pace; thus, complete copies of all the patents issued from 1836 on were not available in print until many years later. The ready availability of inexpensive printed copies of patents did much to increase the usefulness of the technical information contained in the records of the Patent Office and undoubtedly saved many would-be inventors much time and money that would otherwise have been spent on inventing things that had already been patented.

After good, inexpensive printed copies of the drawings became readily available, the originals were no longer needed by the Patent Office in its day-to-day business, and they, like so many other government documents, were transferred to the National Archives for safekeeping. There, they constitute an invaluable part of the documentary record of our nation's technological history.

THE FACSIMILES

Captions for the Documents on the Following Pages

Document 1

WHITNEY'S COTTON GIN, PAT. NO. 72X

Eli Whitney's cotton "engine" ("gin") was patented in 1794. Whitney calculated that a 1-horsepower machine could separate cotton fiber from seeds faster than 50 men working by the old methods. This stimulated a great need for slaves to pick the cotton to keep this simple machine operating at capacity. Whitney's cotton gin, its basic design still in use today, made cotton the staple crop of the South, which by 1808 had increased its exportation twentyfold. The original drawing was destroyed by a fire in the Patent Office in 1836. The drawing shown here was made in 1845 by an artist employed by the Patent Office to replace the one lost in the fire. The "X" after the patent number indicates that it belongs to the series of similarly "restored" patent records.

Document 2

T. BLANCHARD'S TURNING MACHINE, PAT. NO. 3131X

In 1818, Thomas Blanchard designed a special-purpose lathe to produce gunstocks. This invention proved a benchmark in the manufacture of interchangeable precision parts. While Blanchard mulled over how to turn a gunstock, the principle of turning irregular forms from a pattern became clear to him. Models were constructed that worked so well that shoe lasts, spokes of wheels, hat blocks, and any number of consumer goods could be produced as well as gunstocks. This invention far exceeded Blanchard's initial goal of reducing the labor requirements of U.S. armories. The designs for many machine tools, woodworking machines, and even dies for medals now incorporate his ideas. His patent was renewed in 1834 and was acknowledged by Congress to stand among the "first rank of American inventions."

Document 3

HOWE'S SEWING MACHINE, PAT. NO. 4750

Although the performance of the apparatus pictured in Elias Howe's patent drawing left much to be desired, it is considered by historians to be the first practical sewing machine. With a curved, eyepointed needle and a long shuttle that moved horizontally underneath the cloth, it was the first patented lockstitch machine (1846). This sewing machine accomplished 250 stitches per minute and revolutionized the corset-making industry in England. Later improvements, primarily those of Isaac M. Singer, facilitated a revolution of the U.S. garment industry, enabling the production of clothing of uniform good quality in mass quantities at an economical price.

Howe himself manufactured relatively few copies of his patented machine, but after many legal battles he became rich collecting royalties from others, such as Singer, who manufactured and sold improved versions of it.

Documents 4a and 4b

Bell's Telephone, Pat. No. 174,465, Drawing and Oath

Bell's telephone, an apparatus for transmitting speech "by causing electrical undulations similar in form to the vibrations of the air accompanying the said vocal or other sounds," is illustrated in figure 7 of his patent drawing. It almost seems to be an afterthought in this patent, which was otherwise directed to a system for sending a plurality of electric telegraph signals simultaneously over the same wire.

On March 10, 1876, 3 days after this patent was issued, Bell became the first to transmit human speech via telephone. His work culminated in one of the most profitable and most contested of all 19th-century patents. Massive and complex, today's telephone systems enable a person to dial a few numbers and in seconds be in communication with any one of an estimated 600 million telephones worldwide. Telephone lines are essential in transmitting television and radio programs and computer data.

Documents 5a and 5b

Edison's Electric Light, Pat. No. 223,898, Drawing and Claim

Thomas A. Edison did not invent the incandescent lamp, but rather made it practical and commercially viable by incorporating a carbonized cotton filament that would not burn out rapidly and a lamp that could be powered from central stations. It rapidly replaced the gas flame lamp, kerosene-oil lamp, and wax candle, representing a revolution in convenience, economy, and safety in lighting, especially for the home.

Physically, Edison's electric light is the height of simplicity: a carbon filament sealed in a vacuum inside a glass bulb. But conceptually, it represents a radical departure from earlier attempts to "sub-divide" electric light. Where others believed a plurality of electric lights could only be connected in a series, Edison's long-lasting, high-resistance filament permitted him to connect them in parallel. The economies in the associated power generating and distribution system afforded by this arrangement contributed as much to the success of electric lighting as Edison's light source itself.

Documents 6a, 6b, and 6c

Wright's Flying Machine, Pat. No. 821,393, Perspective, Side View, Top View, and Detail

Before the Wright brothers' first successful flight in a powered aircraft in 1903, they made several test flights in unpowered gliders of various designs. Patent number 821,393, applied for in the same year as their first powered flight, covers one of these. Once the best glider design was arrived at, adding a propeller and gasoline engine to it was a relatively simple matter, but it launched man on a revolution in transportation. The following decades saw the rapid growth of this new means of transport, spurred by military needs and spectacular achievements by civil aviation in engines, aircraft design, navigational aids, lighted and well-equipped airports, and communications. Air transportation, both passenger and freight, has assumed the stimulative role in the U.S. economy that railroads played during the previous century, and today production of airplanes has become a major world industry.

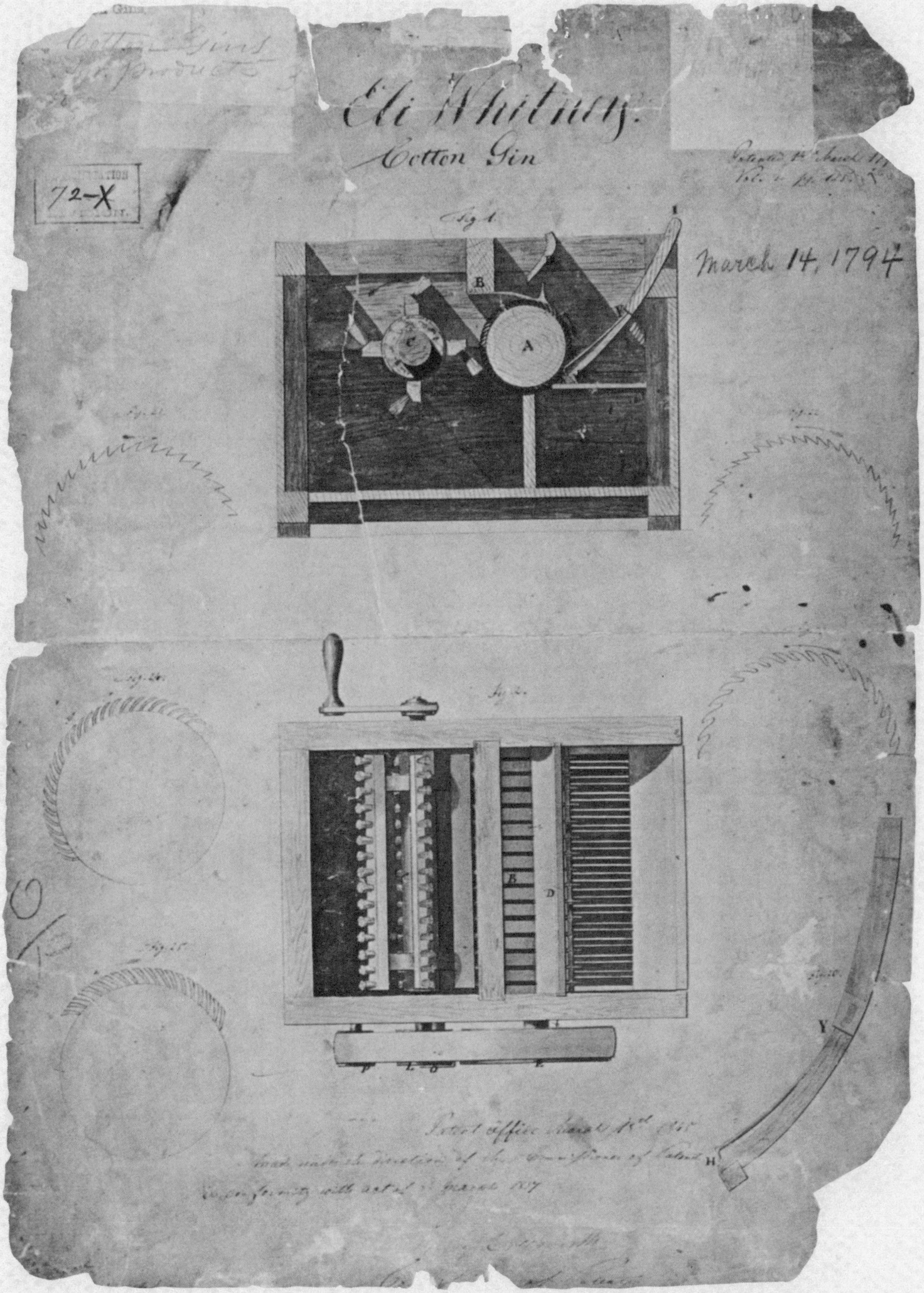

WHITNEY'S COTTON GIN, PAT. NO. 72X *Document 1*

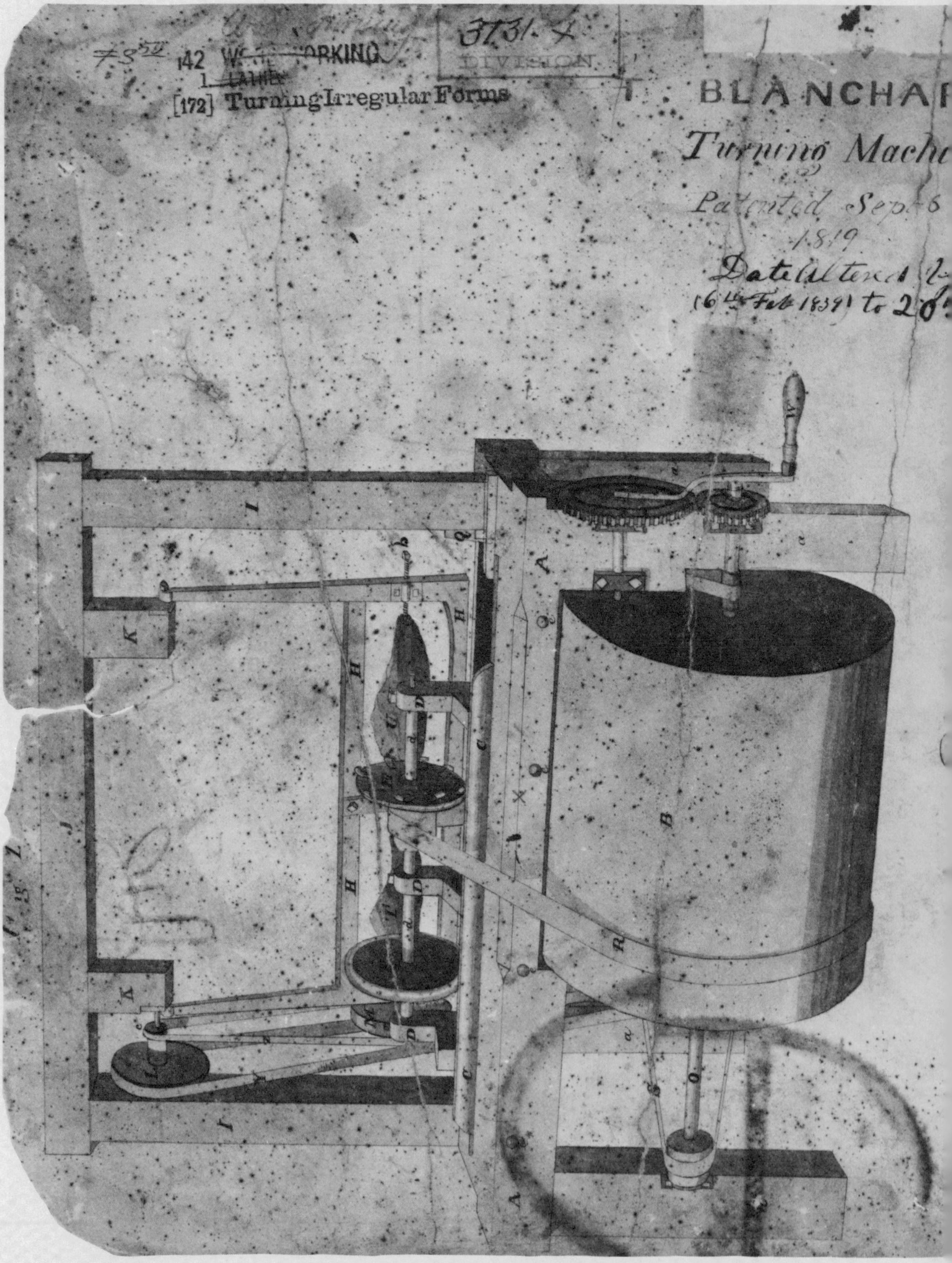

Document 2 T. BLANCHARD'S TURNING MACHINE, PAT. NO. 3131X

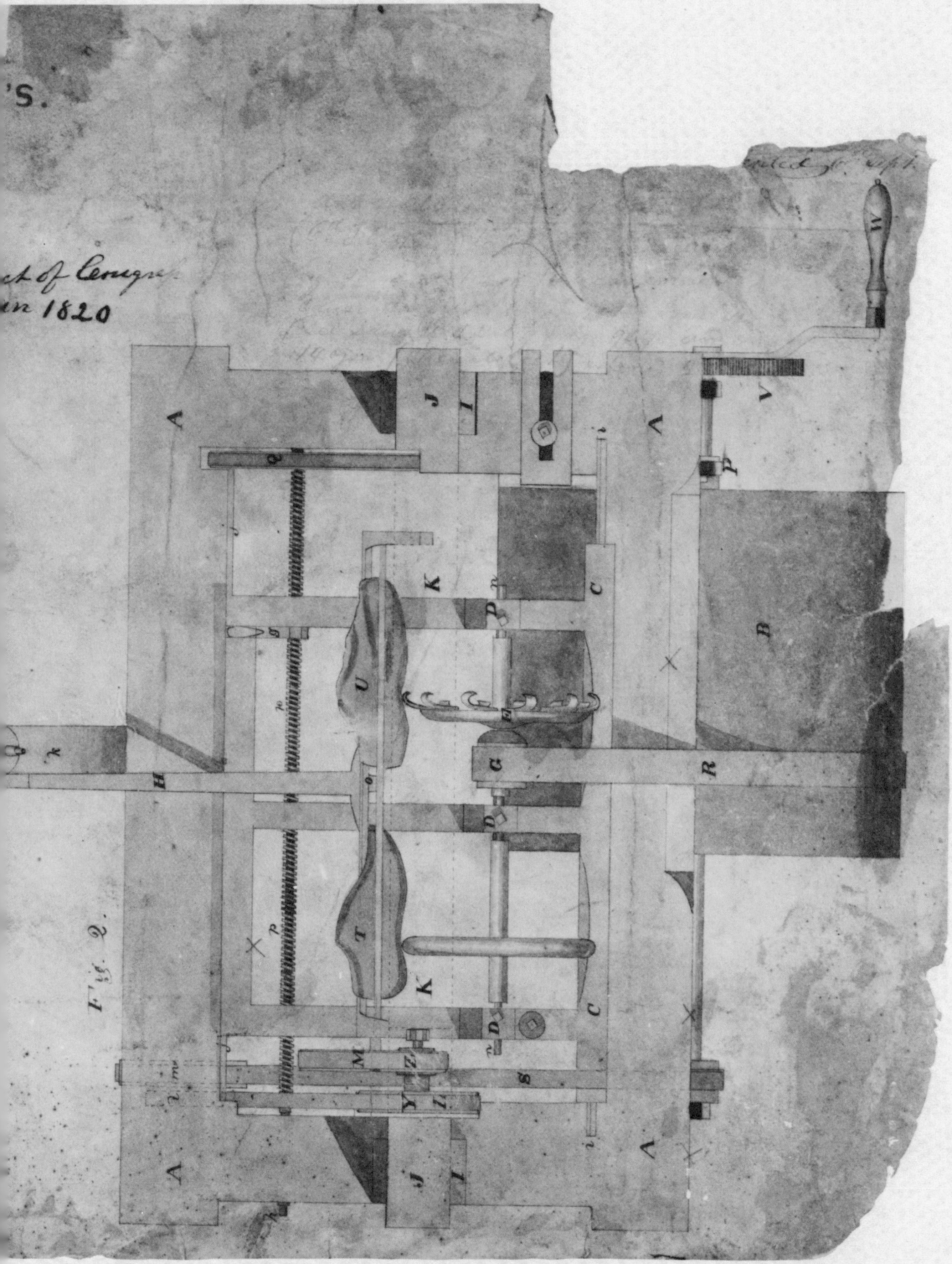

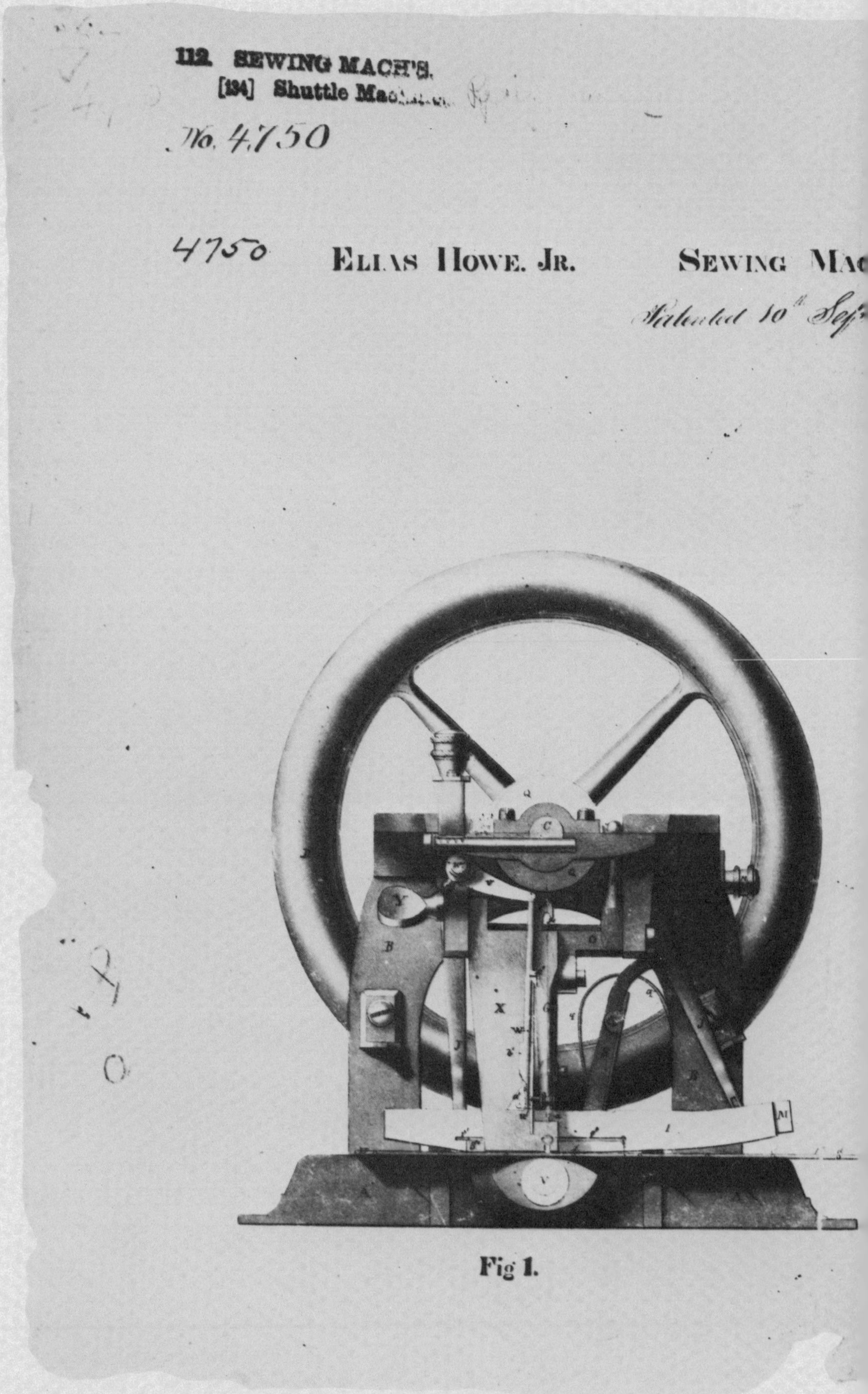

Document 3 HOWE'S SEWING MACHINE, PAT. NO. 4750

Sheet 1 2 Sheets

E.

1846

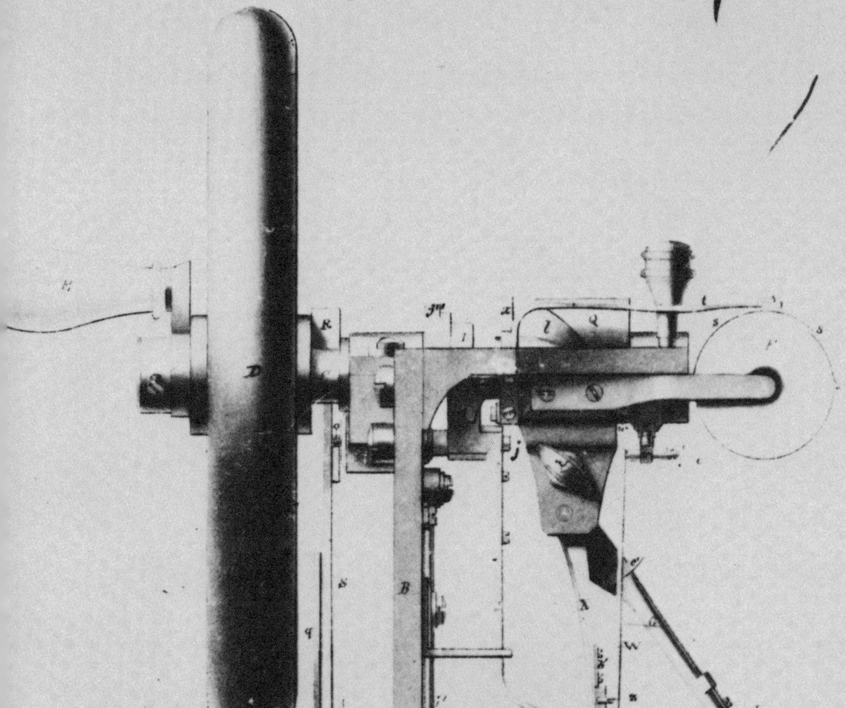

Fig 2.

2 Sheets—Sheet 2.

A. G. BELL.

TELEGRAPHY.

No. 174,465. Patented March 7, 1876.

Fig 6.

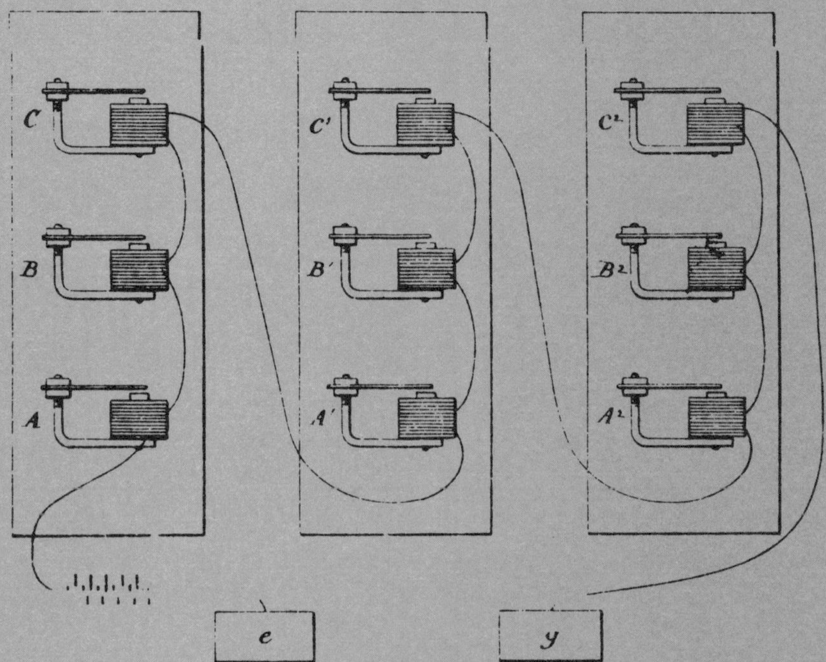

Fig. 7

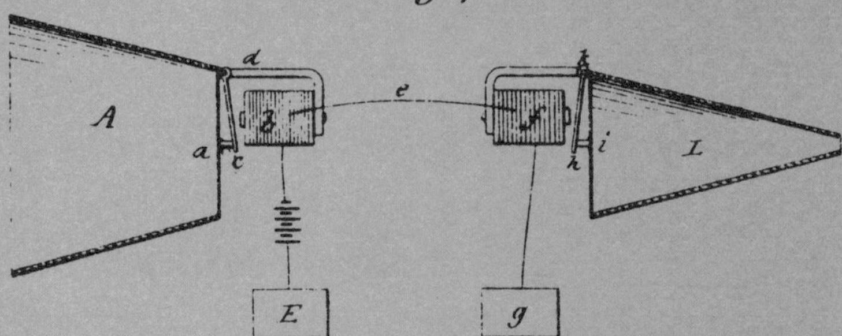

Witnesses

Ewell A. Dick

W. J. Hutchinson

Inventor

A. Graham Bell

by atty Pollok & Bailey

Document 4a BELL'S TELEPHONE, PAT. NO. 174,465, DRAWING

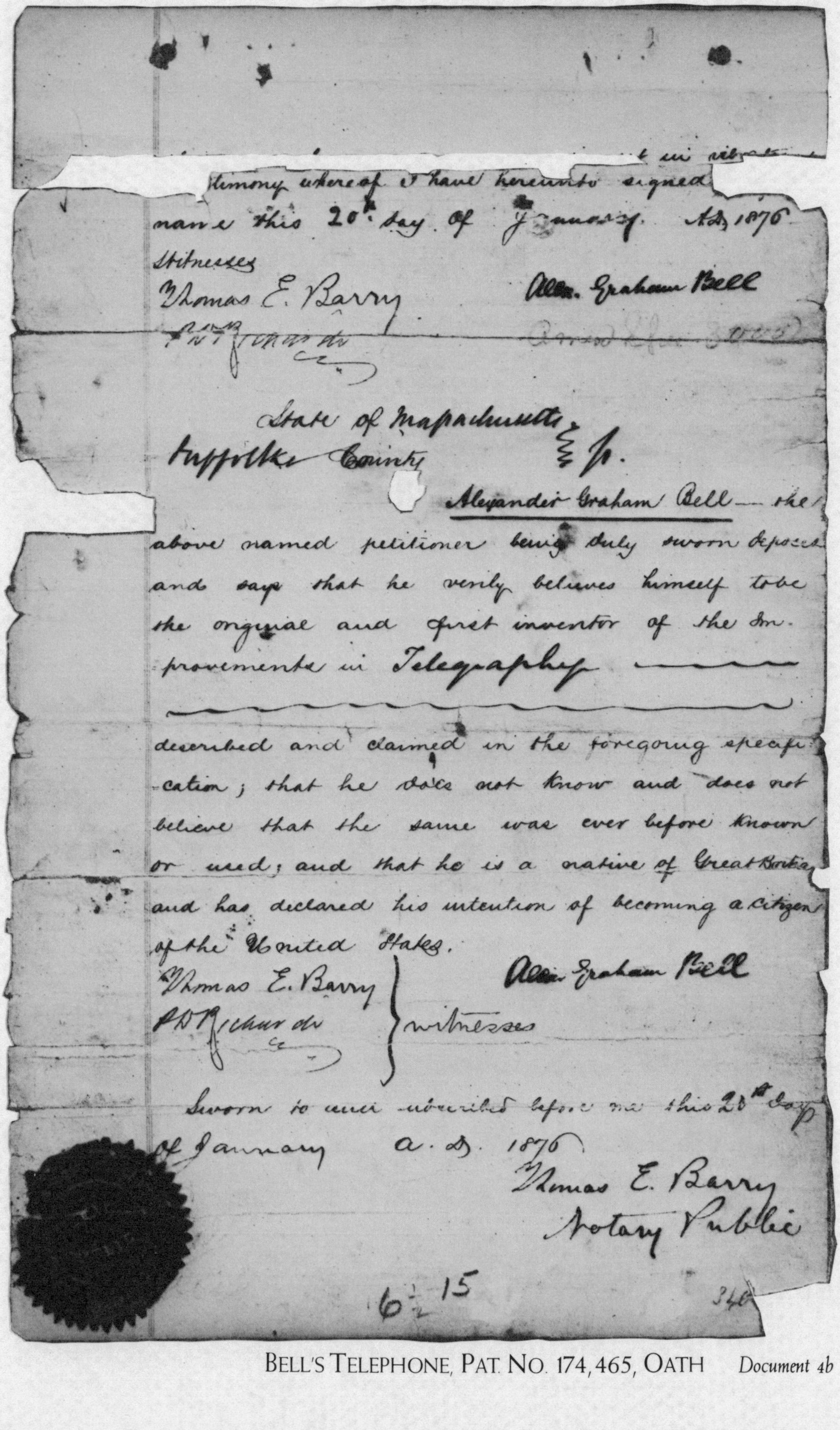

...t in relation...

[In tes]timony whereof I have hereunto signed [my] name this 20th day of January A.D. 1876

Witnesses

Thomas E. Barry — Alex. Graham Bell

P. D. Richards

State of Massachusetts } ss.
Suffolk County }

Alexander Graham Bell the above named petitioner being duly sworn deposes and says that he verily believes himself to be the original and first inventor of the Improvements in Telegraphy described and claimed in the foregoing specification; that he does not know and does not believe that the same was ever before known or used; and that he is a native of Great Britain and has declared his intention of becoming a citizen of the United States.

Alex. Graham Bell

Thomas E. Barry }
P. D. Richards } witnesses

Sworn to and subscribed before me this 20th day of January A.D. 1876

Thomas E. Barry
Notary Public

6½ 15

340

BELL'S TELEPHONE, PAT. NO. 174,465, OATH *Document 4b*

T. A. EDISON.
Electric-Lamp.

No. 223,898. Patented Jan. 27, 1880.

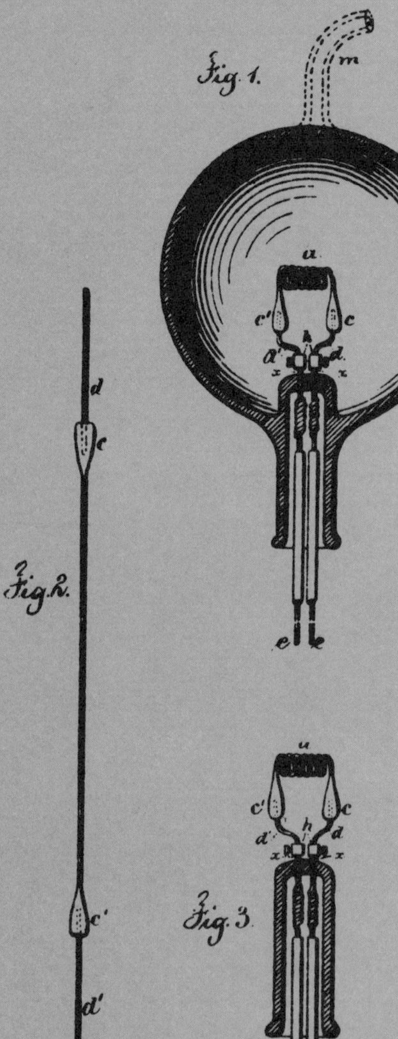

Witnesses
Chas. H. Smith
Geo. T. Pinckney

Inventor
Thomas A. Edison
for Lemuel W. Serrell
atty

Document 5a EDISON'S ELECTRIC LIGHT, PAT. NO. 223,898, DRAWING

before being wound into a spiral.-
Fig. 3 shows the spiral after carbonization ready to have a bulb blown over it.-

I claim as my Invention,

First. An electric lamp for giving light by incandescence, consisting of a filament of carbon of high resistance, made as described, and secured to metallic wires as set forth.-

Second. The combination of carbon filaments within a receiver made entirely of glass through which the leading wires pass and from which receiver the air is exhausted for the purposes set forth.-

Third. A coiled carbon filament or strip arranged in such a manner that only a portion of the surface of such carbon conductor shall radiate light as set forth.-

Substitute "B" per amdt. Dec 16. 79

Fourth. The method herein described of securing the platina contact-wires to the carbon filament and carbonizing of the whole in a closed chamber, substantially as set forth.-

PATENTED FILE

Signed by me this 1st day of November A.D. 1879.

Witnesses:
S. L. Griffin
John. F. Randolph.

Thomas A Edison

8

No. 821,393. PATENTED MAY 22, 1906.

O. & W. WRIGHT.

FLYING MACHINE.

APPLICATION FILED MAR. 23, 1903.

3 SHEETS—SHEET 1.

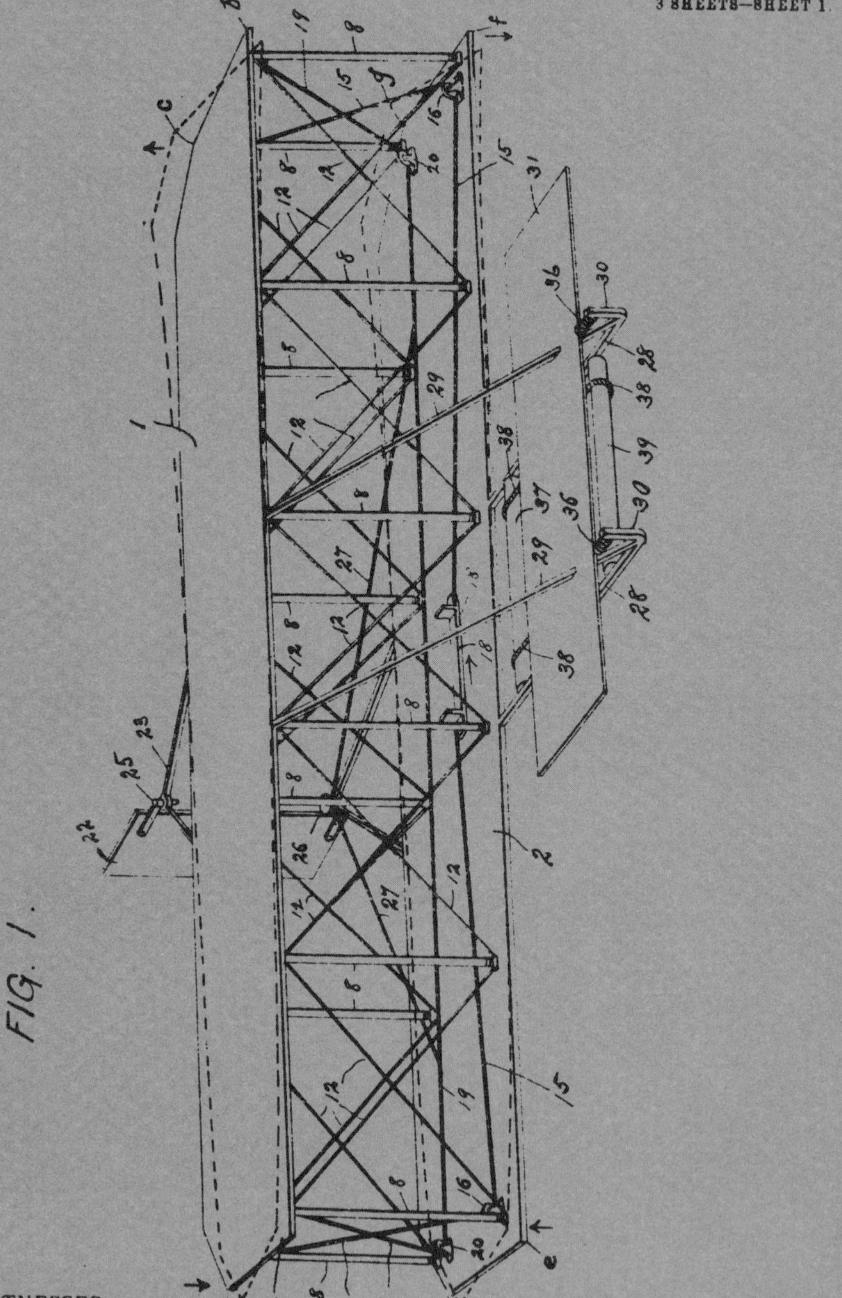

WITNESSES:
William F. Bauer.
Irvine Miller.

INVENTORS.
Orville Wright
Wilbur Wright
BY H. A. Toulmin,
ATTORNEY.

Document 6a WRIGHT'S FLYING MACHINE, PAT. NO. 821,393, PERSPECTIVE, SIDE VIEW

No. 821,393. PATENTED MAY 22, 1906.

O. & W. WRIGHT.

FLYING MACHINE.

APPLICATION FILED MAR. 23, 1903.

3 SHEETS-SHEET 2.

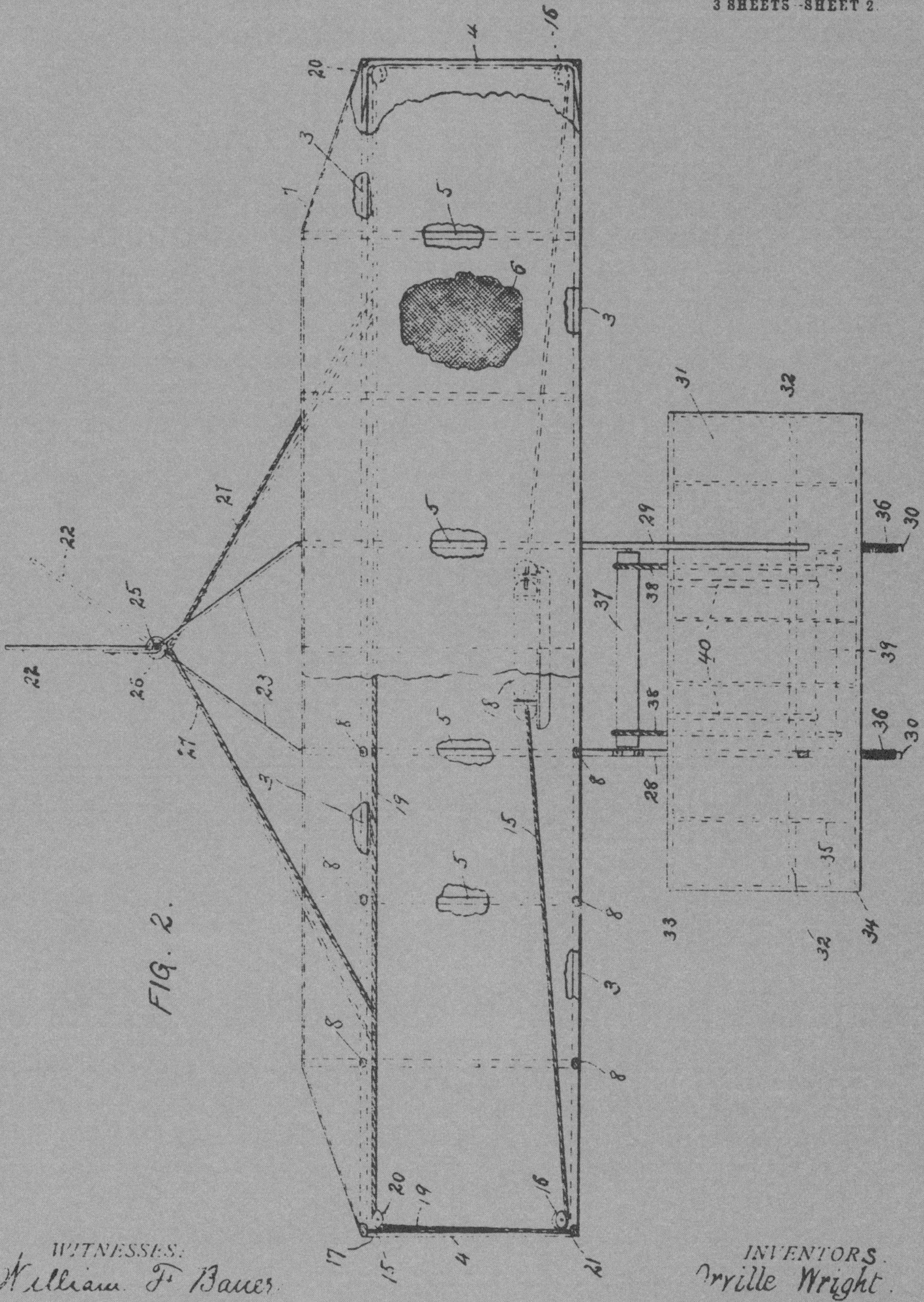

FIG. 2.

WITNESSES:

William F. Bauer

Irvine Miller

INVENTORS.

Orville Wright.

Wilbur Wright.

BY

ATTORNEY.

WRIGHT'S FLYING MACHINE, PAT. NO. 821,393, PERSPECTIVE, TOP VIEW *Document 6b*

No. 821,393. PATENTED MAY 22, 1906.

O. & W. WRIGHT.

FLYING MACHINE.

APPLICATION FILED MAR. 23, 1903.

3 SHEETS—SHEET 3.

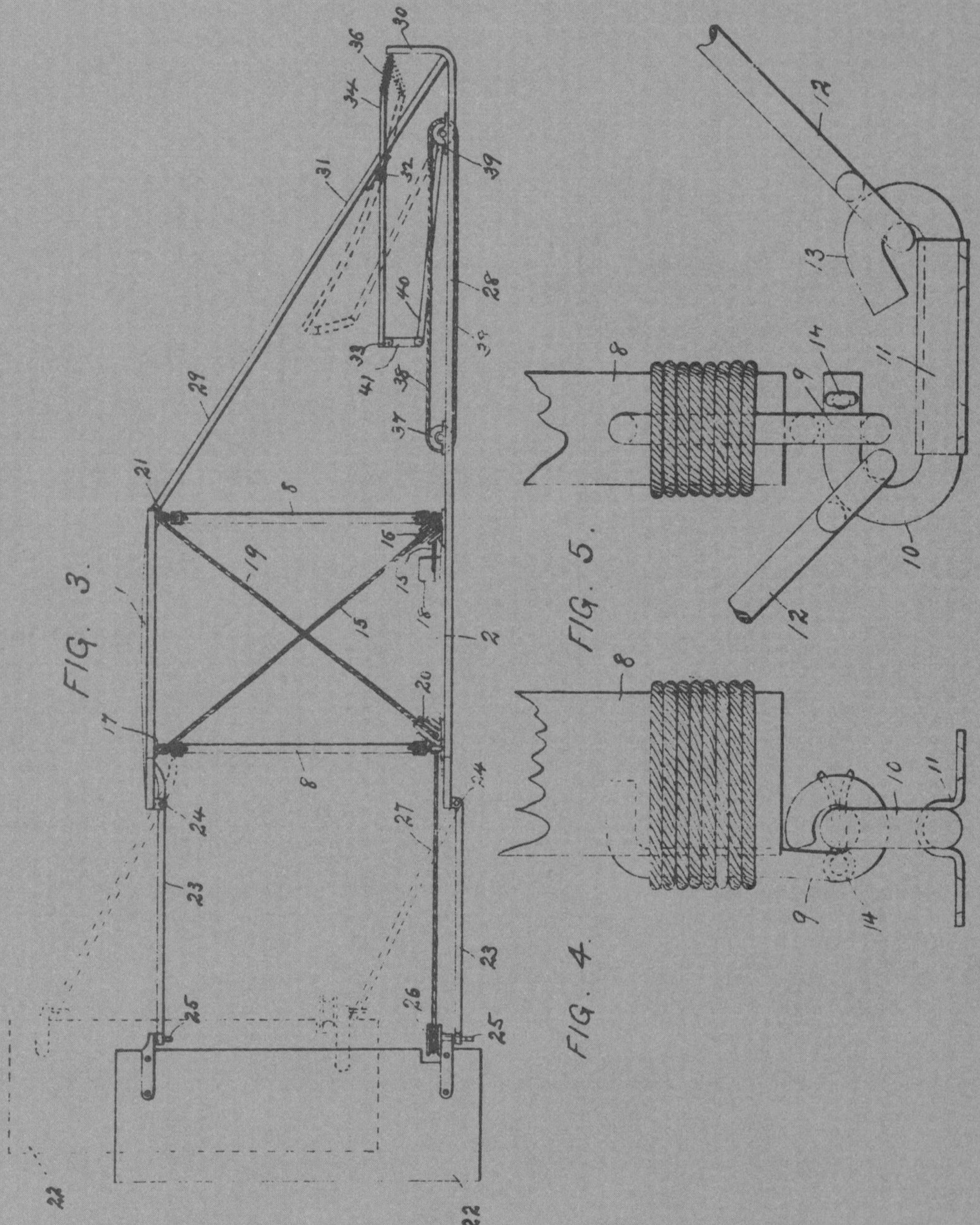

WITNESSES:

William F. Bauer.

Jerome Miller.

INVENTORS.

Orville Wright.

Wilbur Wright.

BY H. A. Toulmin,

ATTORNEY.

Document 6c WRIGHT'S FLYING MACHINE, PAT. NO. 821,393, DETAIL